Amazing Animal Senses!

by Caroline Hutchinson

Table of Contents

What Senses Do Animals Have?

People have five senses. Animals have five senses, too.

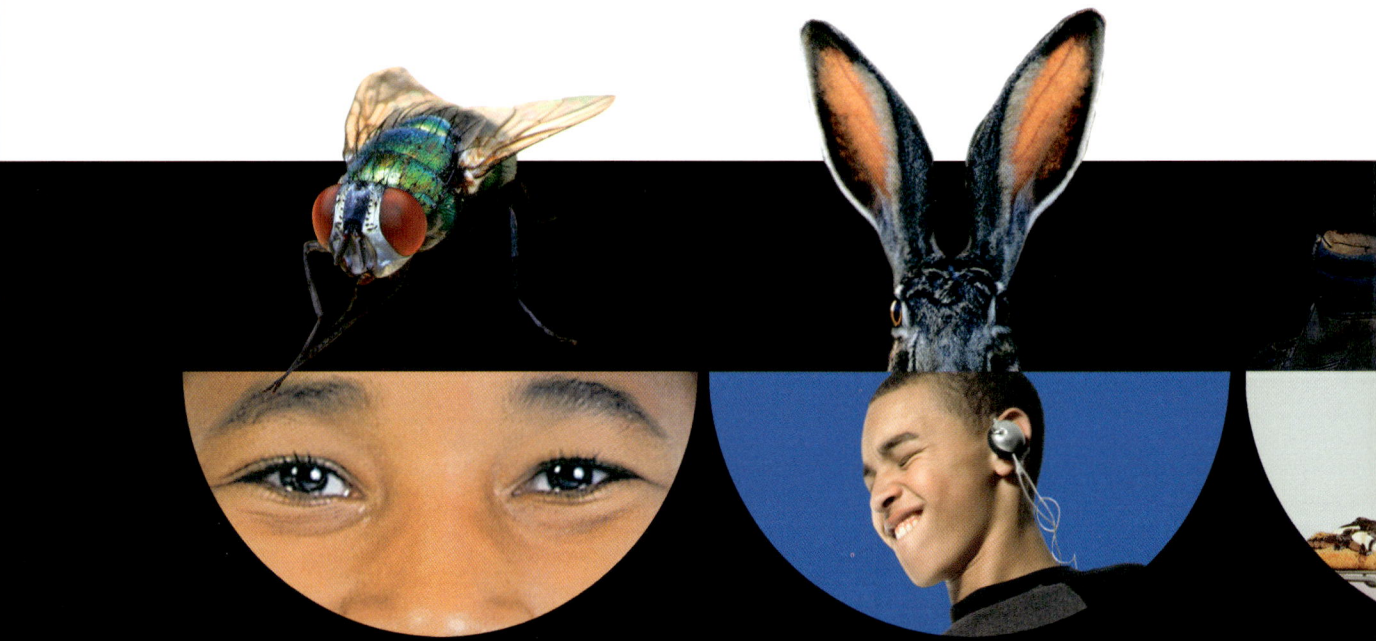

Like people, animals use their senses to find food and stay safe. Some animals have amazing senses.

▲ smell   ▲ taste   ▲ touch

What Animals Have an Amazing Sense of Sight?

Some animals see well in the dark. Cats can see well in the dark.

An owl can see well in the dark, too. The owl's big eyes let in more light. An owl can't move its big eyes very much, so an owl can turn its head around.

◀ This owl can stay safe by seeing what is behind it.

Look at this animal. When one eye is looking up, the other eye can look down. When one eye is looking to the right, the other eye can look to the left.

▲ A chameleon can see food or danger in any direction

Some animals have many eyes. Having many eyes helps these animals find food and stay safe.

▲ A scallop can have as many as one hundred eyes.

▼ This spider has eight eyes.

▲ This jellyfish has twenty-four eyes!

What Animals Have an Amazing Sense of Hearing?

The elephant's big ears help it stay safe. The elephant can hear animals that are near and animals that are far away.

The elephant can also ▶ use these big ears as fans when the weather is very hot.

Can you find the snake's ears? A snake does not have ear holes, but it can hear. A snake uses its bones to hear.

◄ This snake does not have ears on the sides of its head.

Some animals can find food and stay safe in very dark places. These animals make sounds and listen for the echo.

A bat finds food ▶ and stays safe in dark caves.

A dolphin uses ▶ sounds to find food and stay safe under water.

A cricket uses its legs
to hear. The cricket can
find food and other crickets!

This cricket ▶
has ears on
its front legs.

What Animals Have an Amazing Sense of Smell or Taste?

A wet nose helps the dog have an amazing sense of smell. Some animals have an amazing sense of smell because of their long noses.

◀ The aardvark has a very long nose.

A black bear uses its nose and its mouth to smell food and danger. Some people say that black bears have the best sense of smell!

A black bear uses its nose to find food and stay safe. ▼

▲ The shark smells tiny

The tiny bumps on your tongue help you taste your food. Some animals have many more bumps on their tongues.

Your tongue has ▶ about 9,000 tiny bumps.

The rabbit, catfish, ▶ and pig have a better sense of taste than you do.

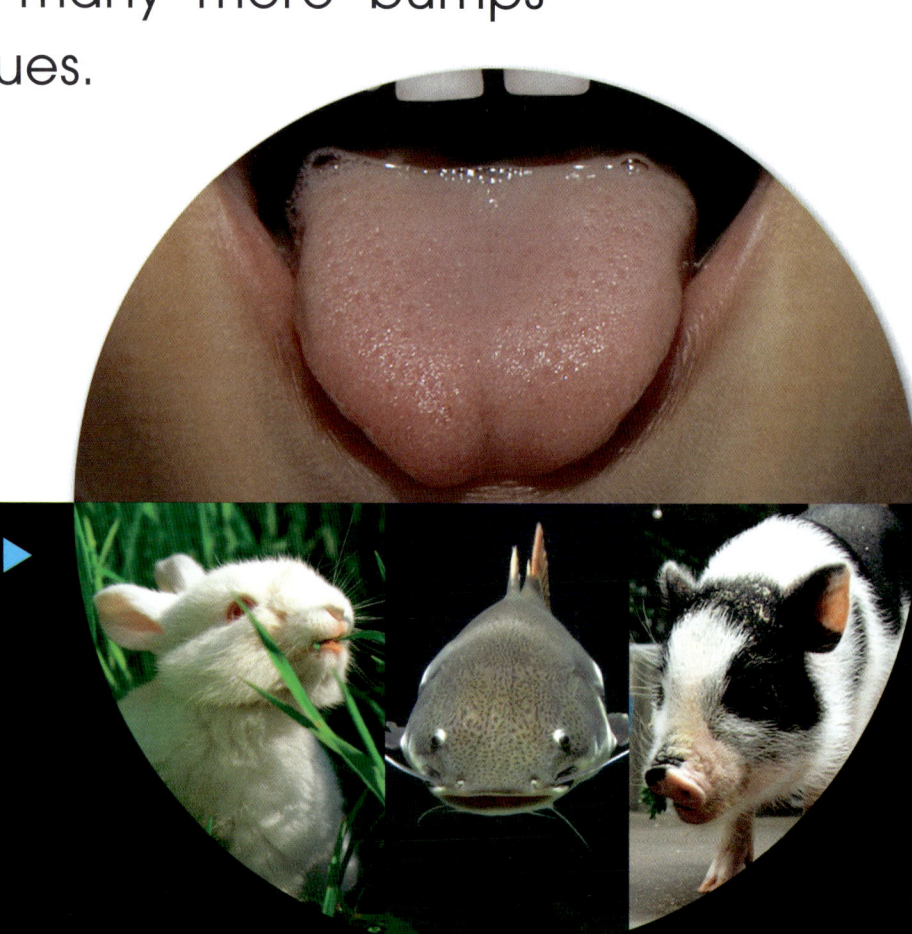

Some insects use their feet to taste! Would you want to use your feet to eat?

This butterfly ▶ eats with its feet.

People and animals have the same five senses. Some animals have amazing senses!